copyright © 2020 MAX DIN

All rights reserved . No part of this book may be reproduced or used in any manner without written permission of the copyright owner , except for the use of quotations and other non commercial uses permitted by copyright law .

SPACE
Coloring Book

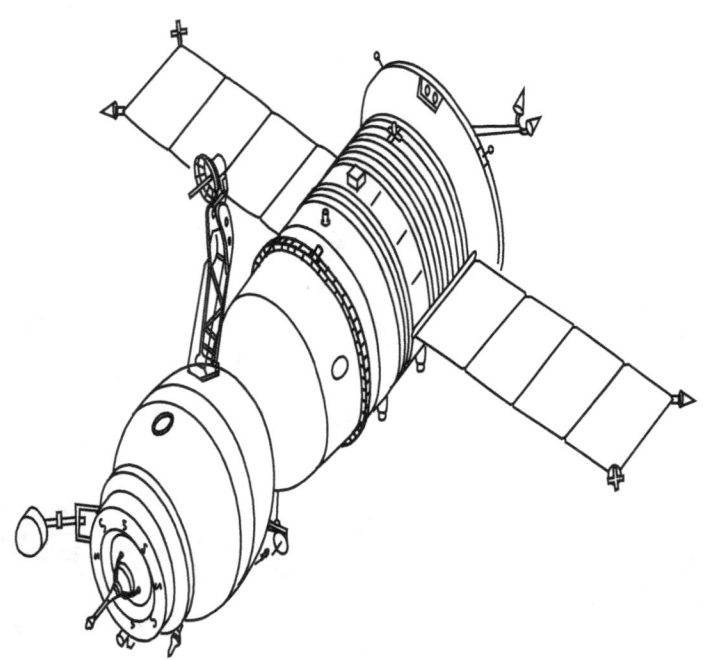

www.vecteezy.com

www.freepik.com

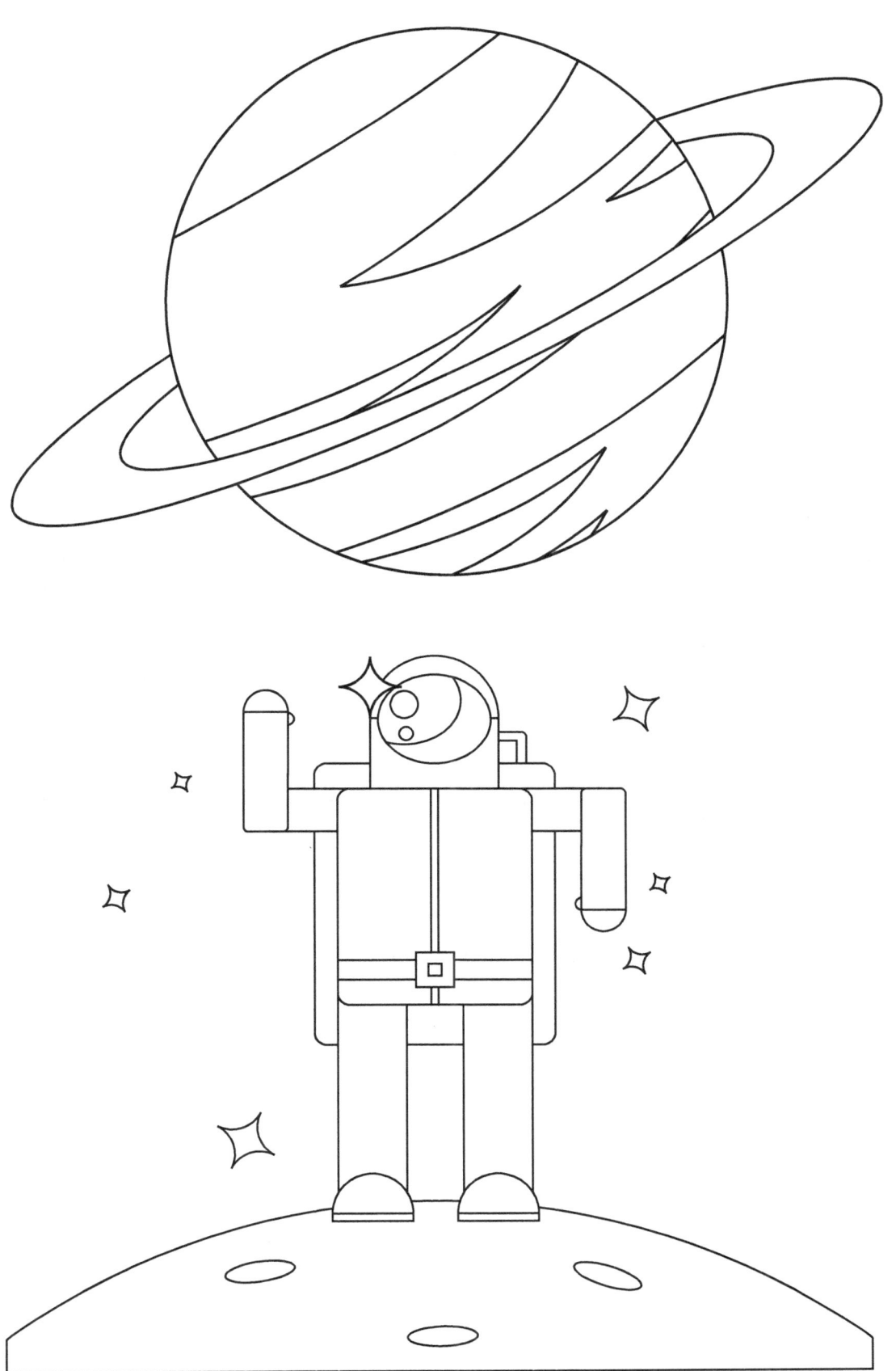

Astronauts use radios to stay in communication while in space, since radio waves can still be sent and received.

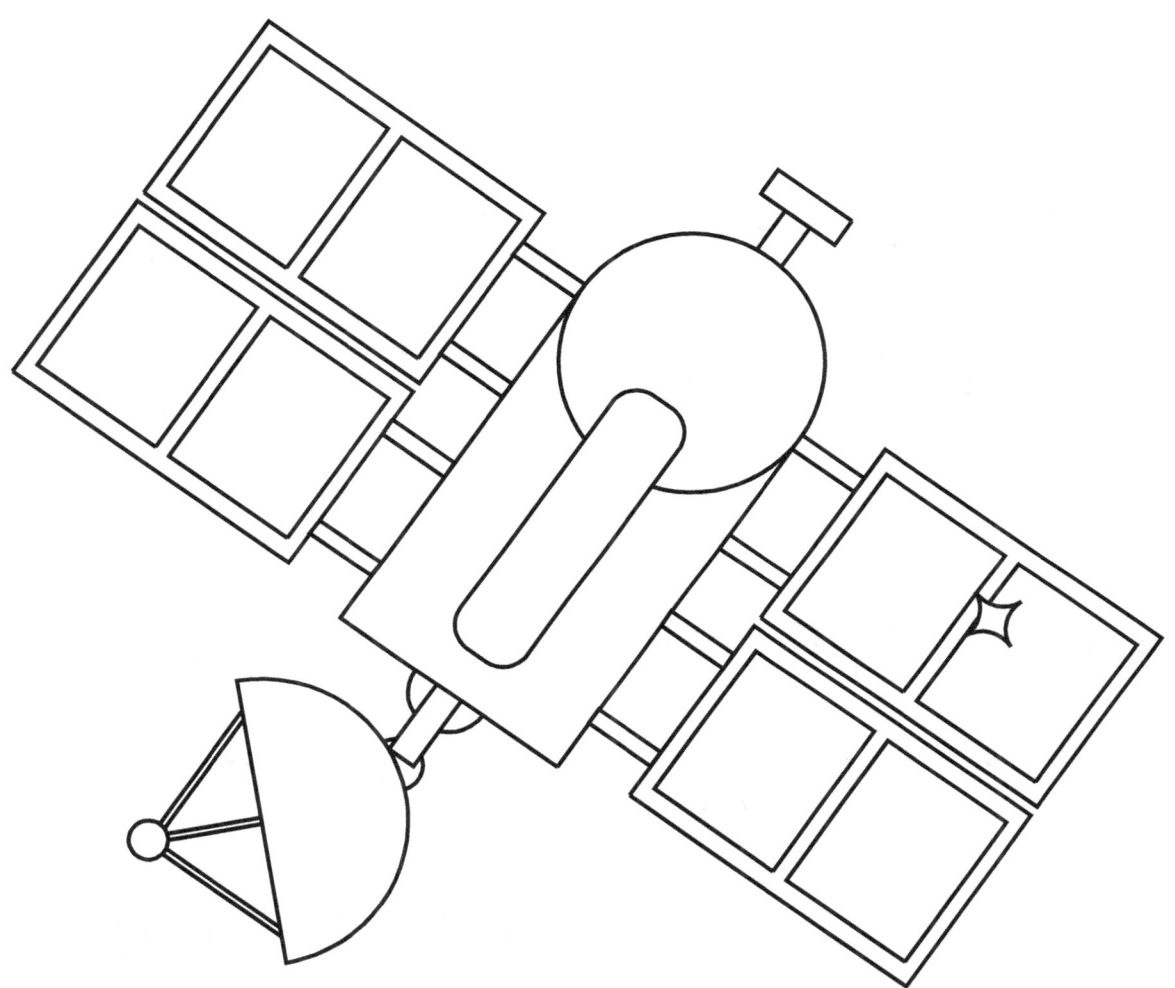

The Milky Way, to estimate. That number is between 200-400 billion stars and there are estimated to be billions of galaxies so the stars in space really are completely uncountable.

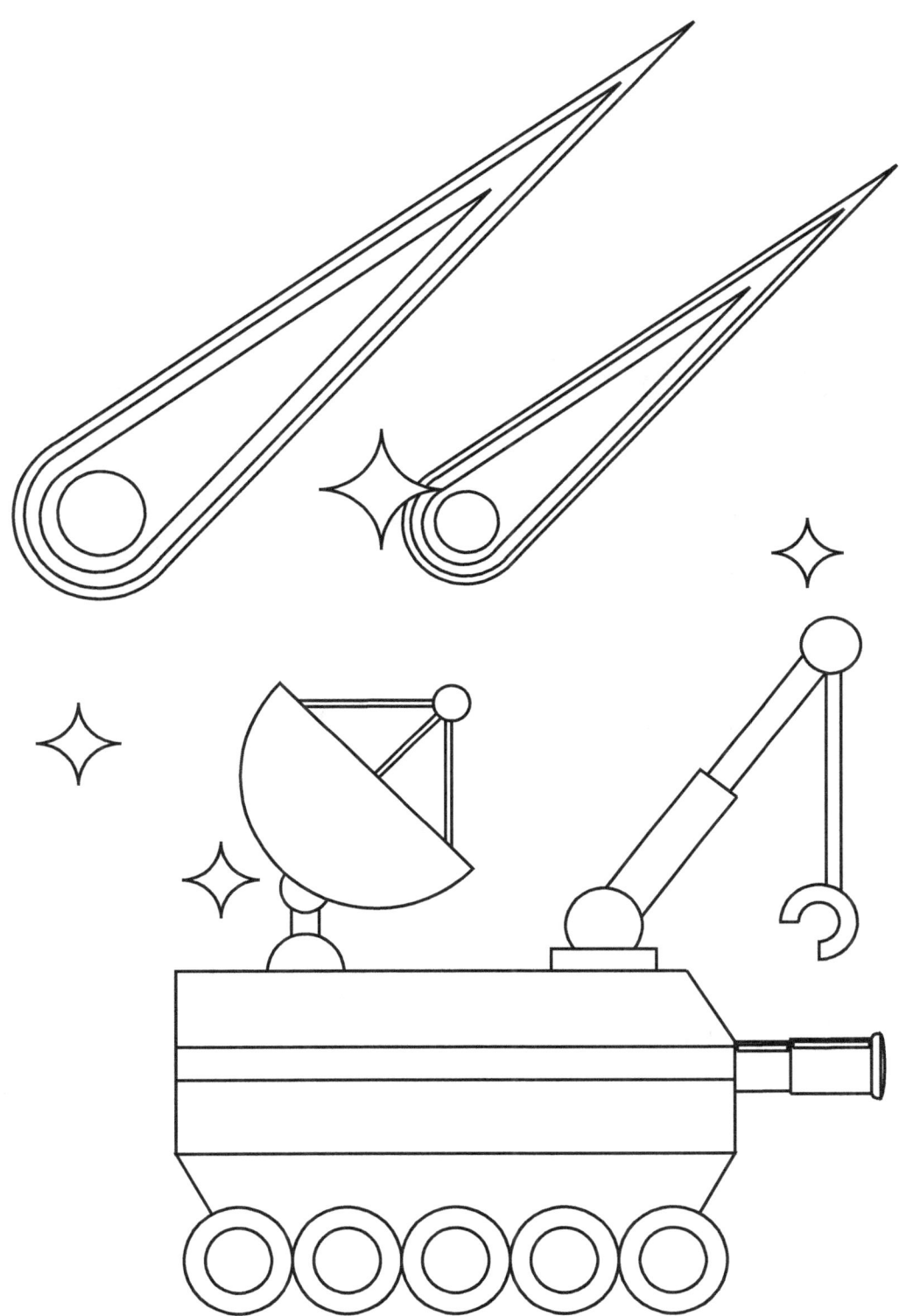

An asteroid about the size of a car enters Earth's atmosphere roughly once a year but it burns up before it reaches us. Phew!

There is no atmosphere in space, which means that sound has no medium or way to travel to be heard.

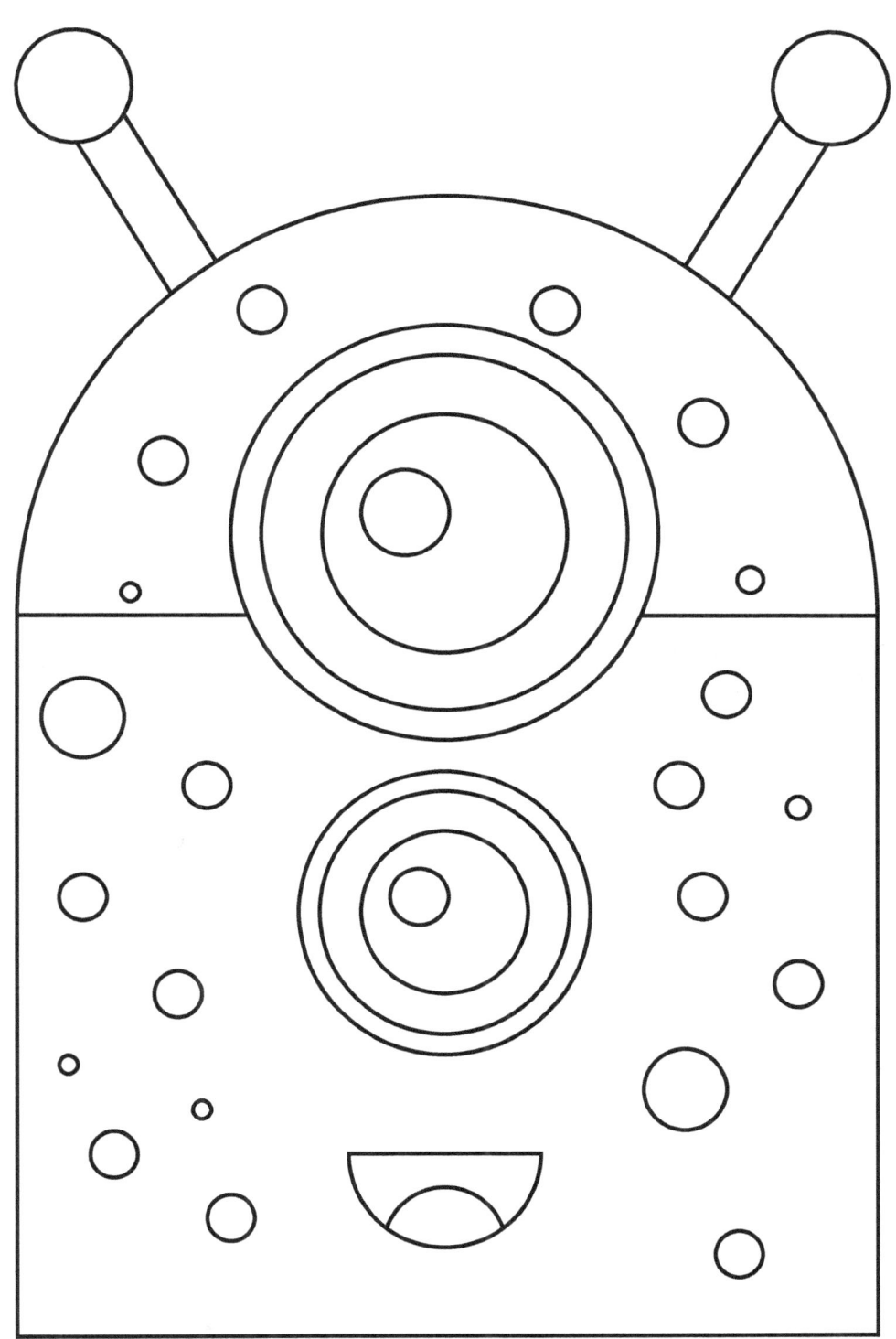

The hottest planet in our solar system is 450° C

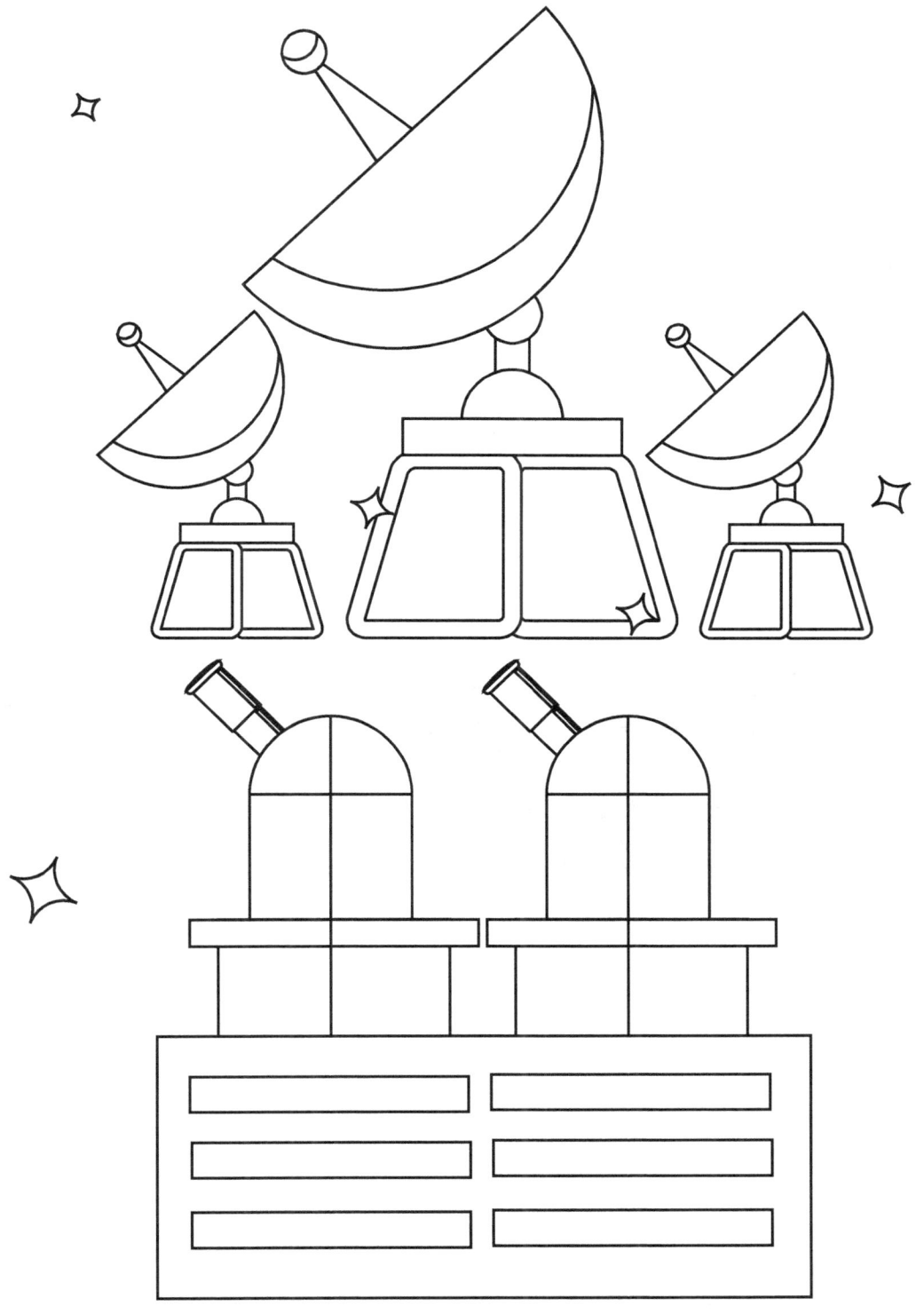

A full NASA space suit costs $12,000,000.

Venus is the hottest planet in the solar system and has an average surface temperature of around 450° C. Interestingly

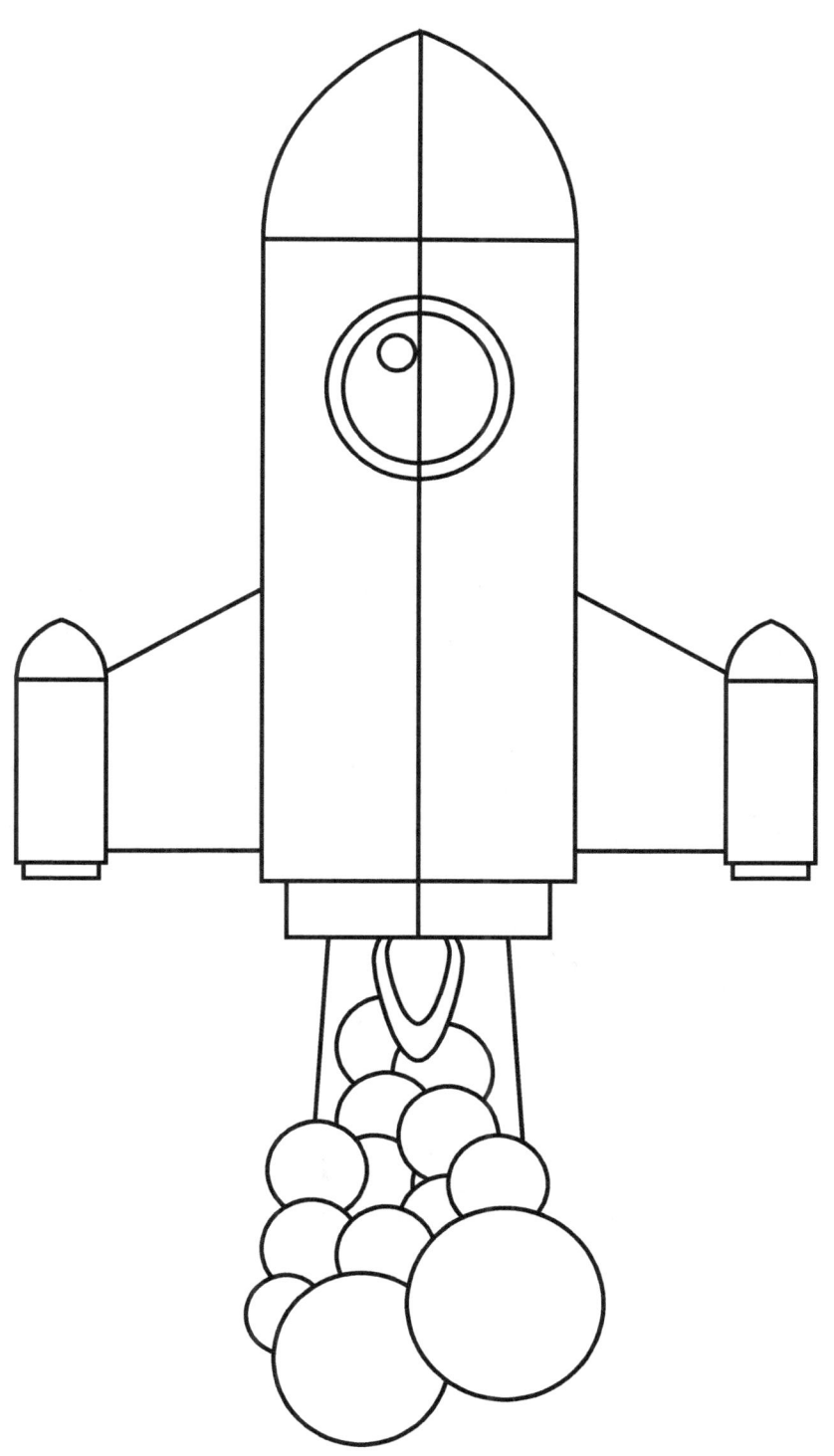

If you could fly a plane to Pluto, the trip would take more than 800 years!

Venus is not the closest planet to the Sun - Mercury is closer but because Mercury has no atmosphere to regulate temperature it has a very large temperature fluctuation.

You wouldn't be able to walk on Jupiter, Saturn, Uranus or Neptune because they have no solid surface!

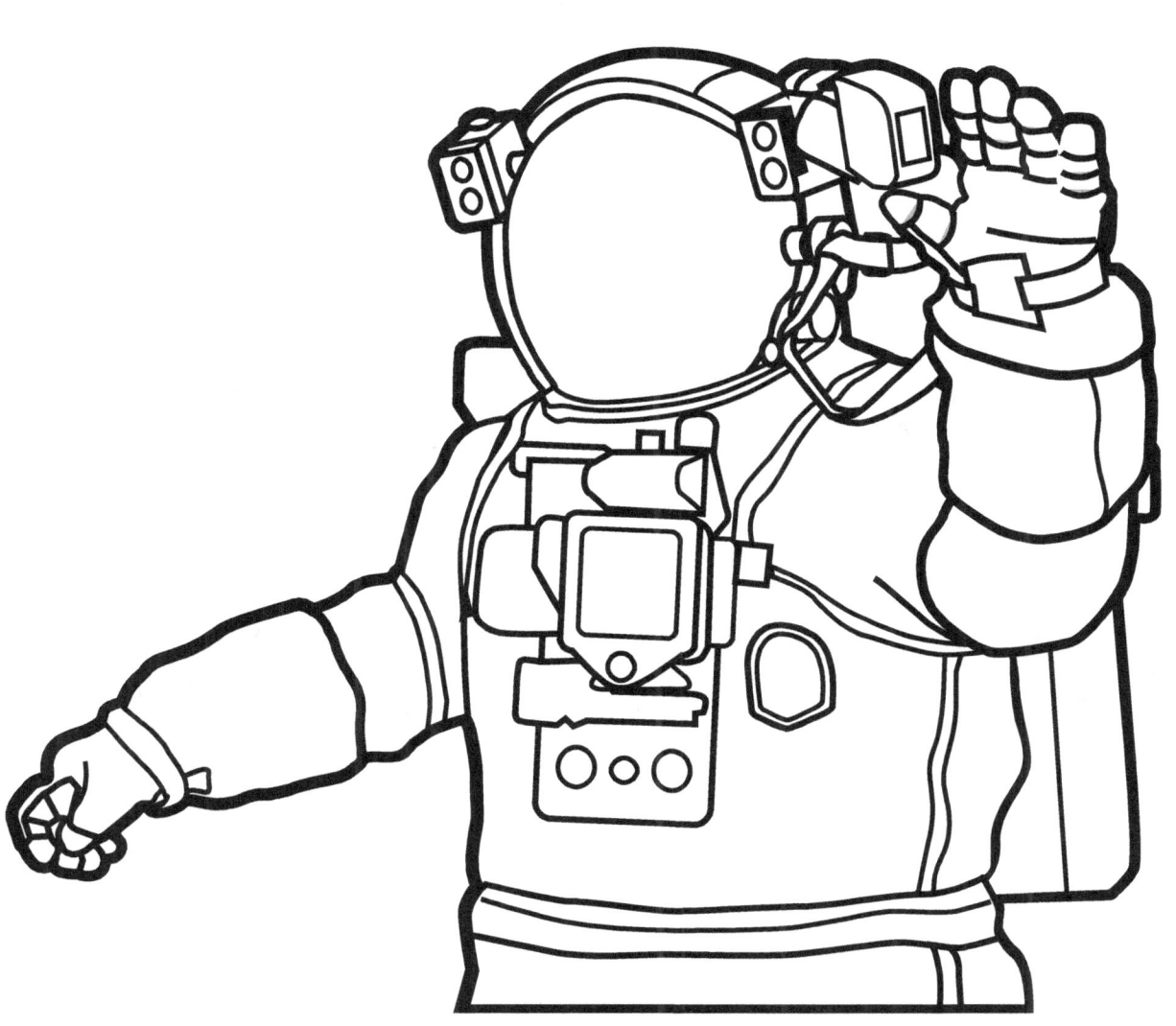

One day on Venus is longer than one year.

The orbit of Venus around the Sun is 225 Earth days.

If two pieces of the same type of metal touch in space they will permanently bond..

There is floating water in space.

The largest known asteroid is 965 km (600 mi) wide..

While the entire suit costs a cool $12m, 70% of that cost is for the backpack and control module.

Neutron stars can spin 600 times per second.

Neutron stars are the densest and tiniest stars in the known universe.

The footprints on the Moon will be there for 100 million years.

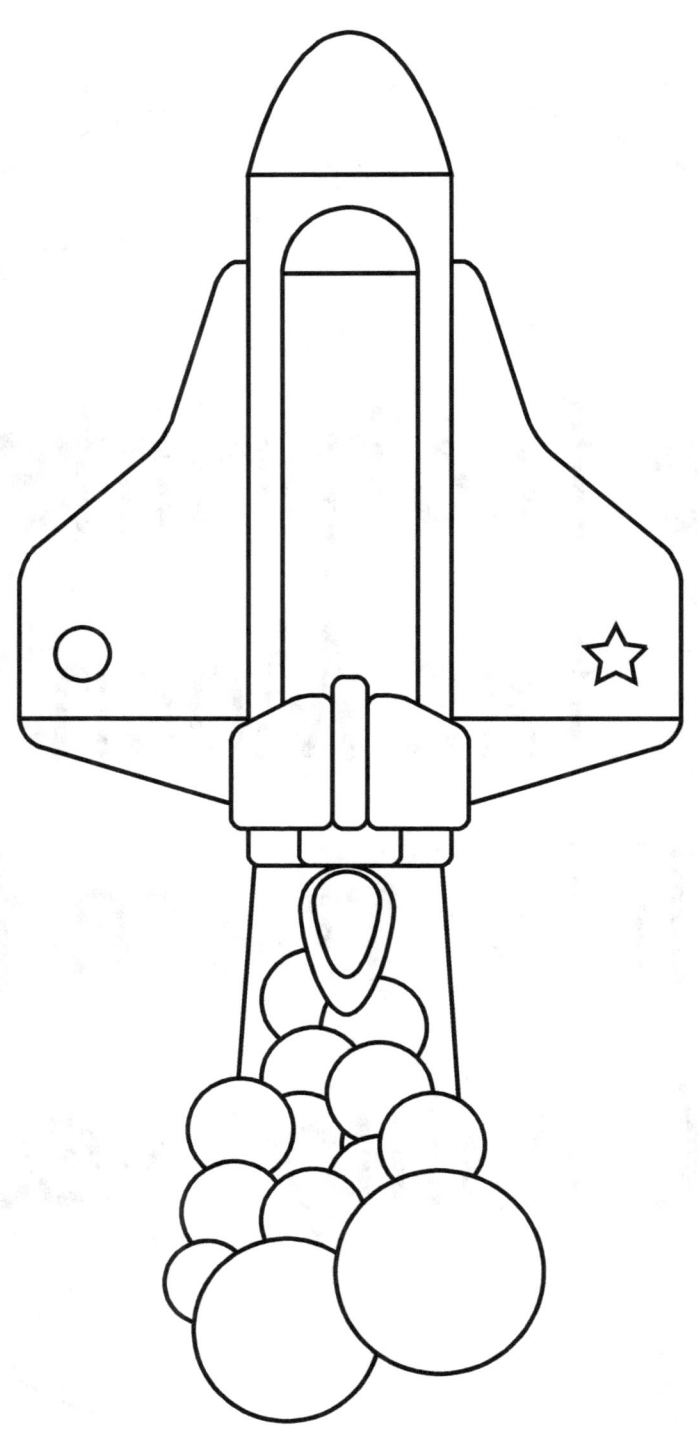

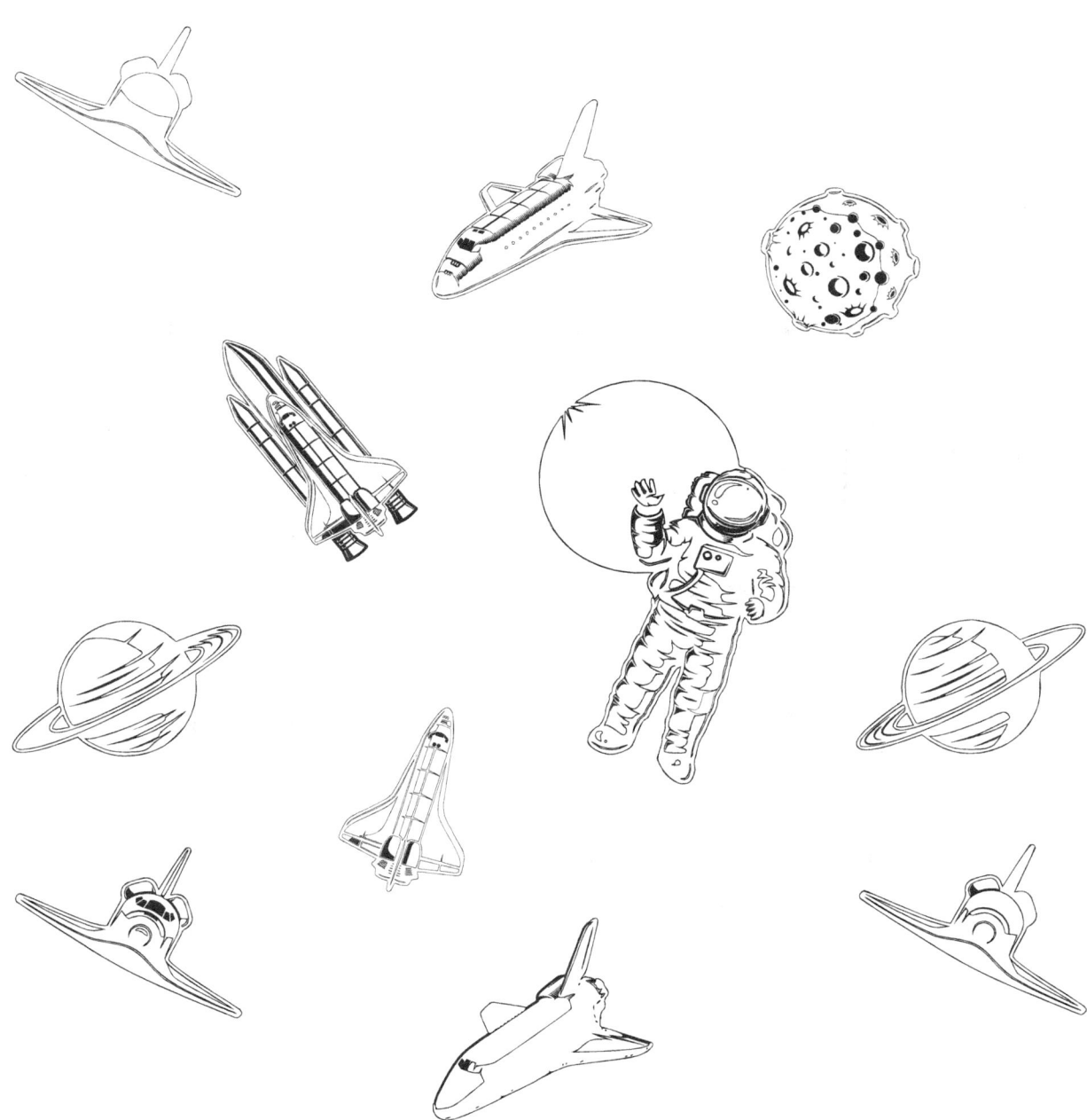

The Moon has no atmosphere, which means there is no wind to erode the surface and no water to wash the footprints away.

The Sun's mass takes up 99.86% of the solar system.

If you could fly a plane to Pluto, the trip would take more than 800 years!

A full NASA space suit costs $12,000,000.

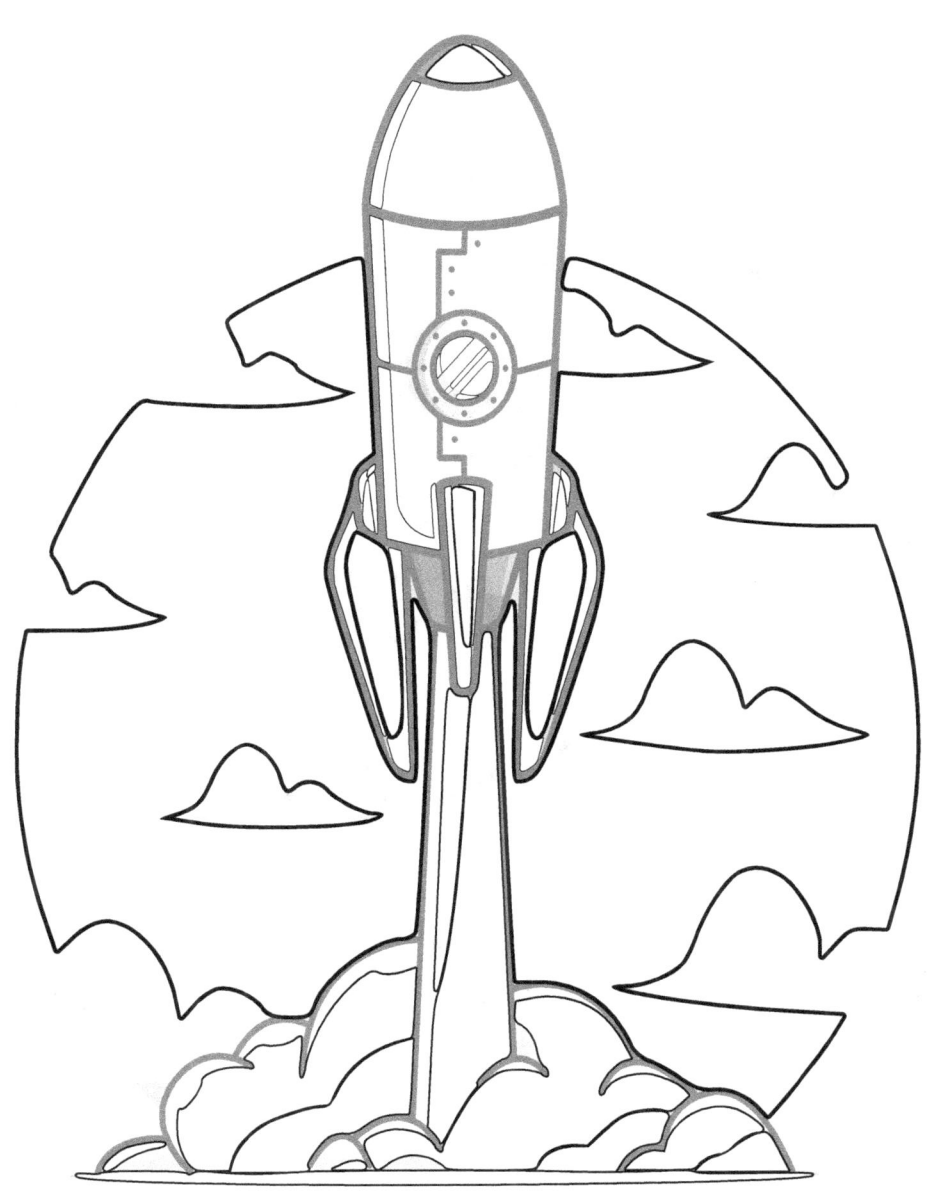

An asteroid about the size of a car enters Earth's atmosphere roughly once a year but it burns up before it reaches us. Phew!

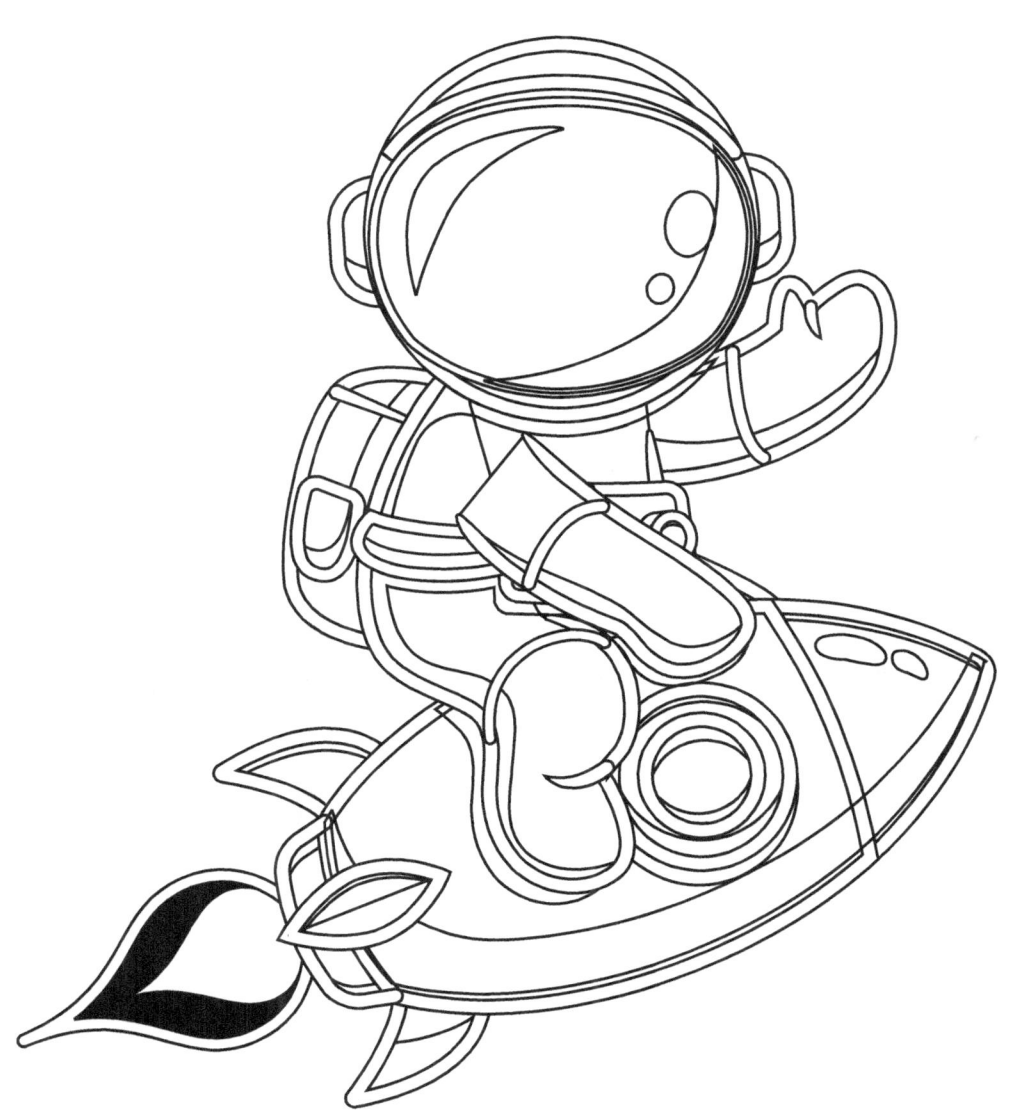

The highest mountain known to man is on an asteroid called Vesta.

There is a volcano on Mars three times the size of Everest.

At 600 km wide and 21 km high, Olympus Mons is a volcano on Mars that may still be active

Comets are leftovers from
the creation of our solar system
about 4.5 billion years ago
they consist of sand,
ice and carbon dioxide.

You wouldn't be able to walk on Jupiter, Saturn, Uranus or Neptune because they have no solid surface!

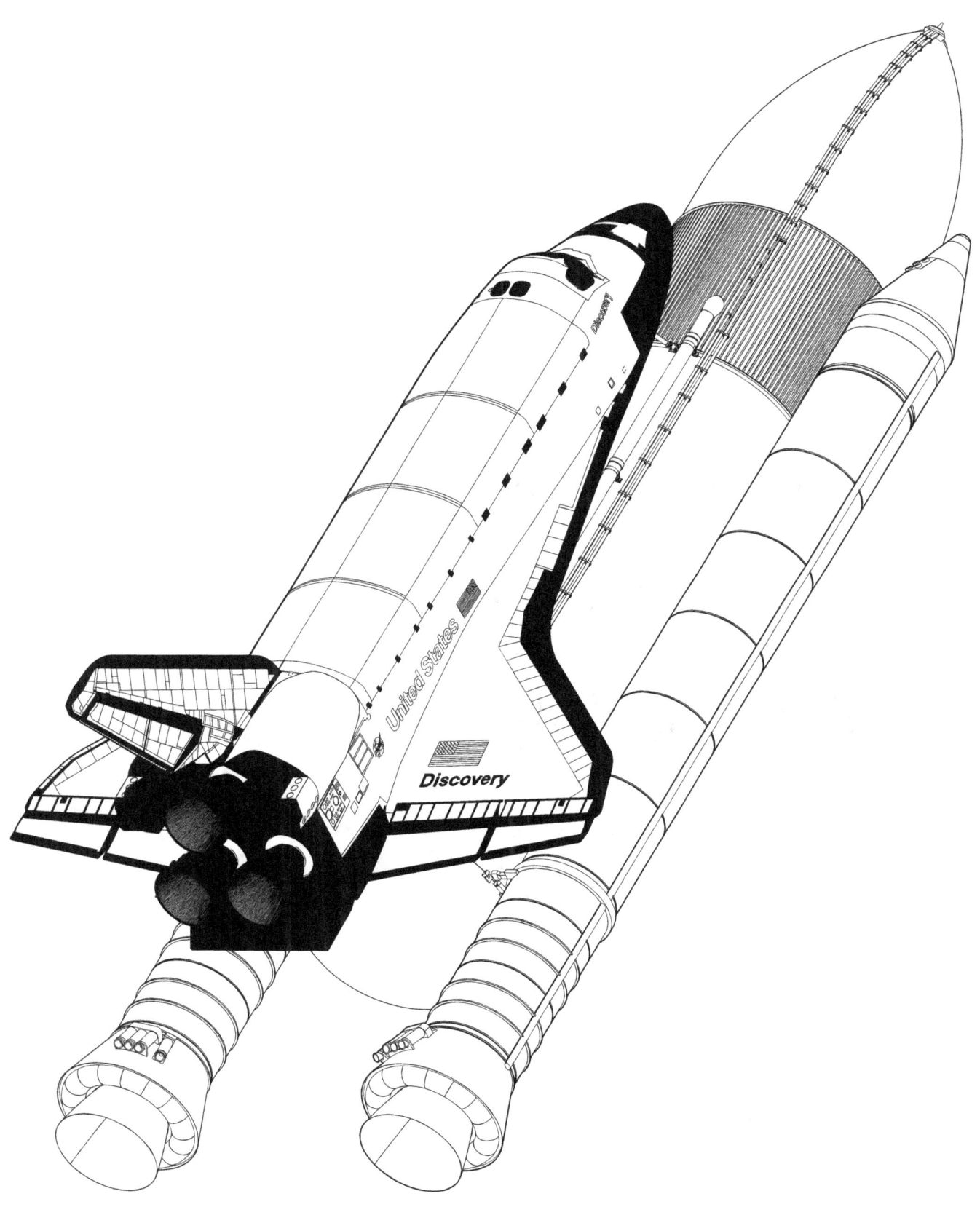

One million Earths could fit inside the sun and the sun is considered an average-size star.

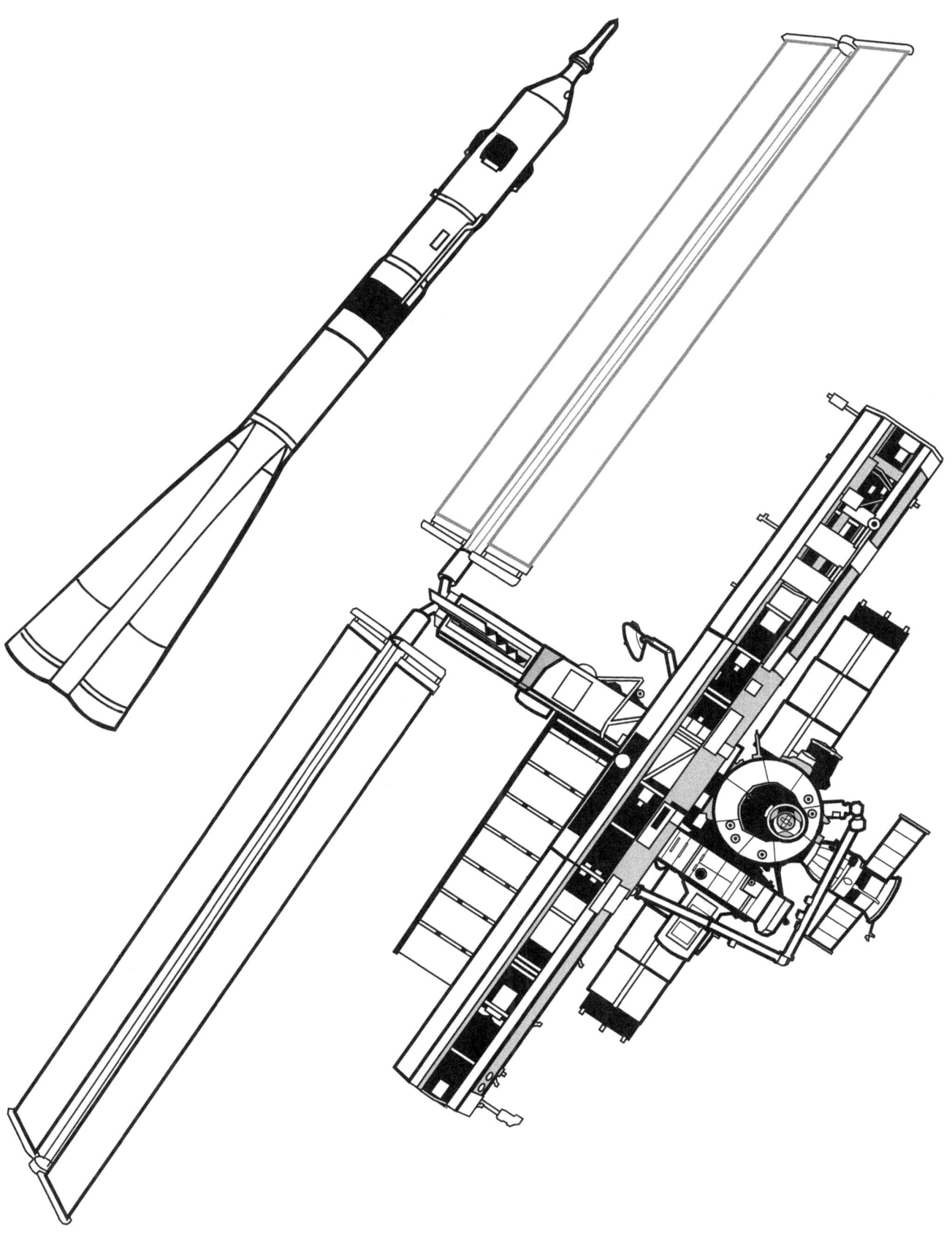

There are more stars in the universe than grains of sand on all the beaches on Earth. That's at least a billion trillion!

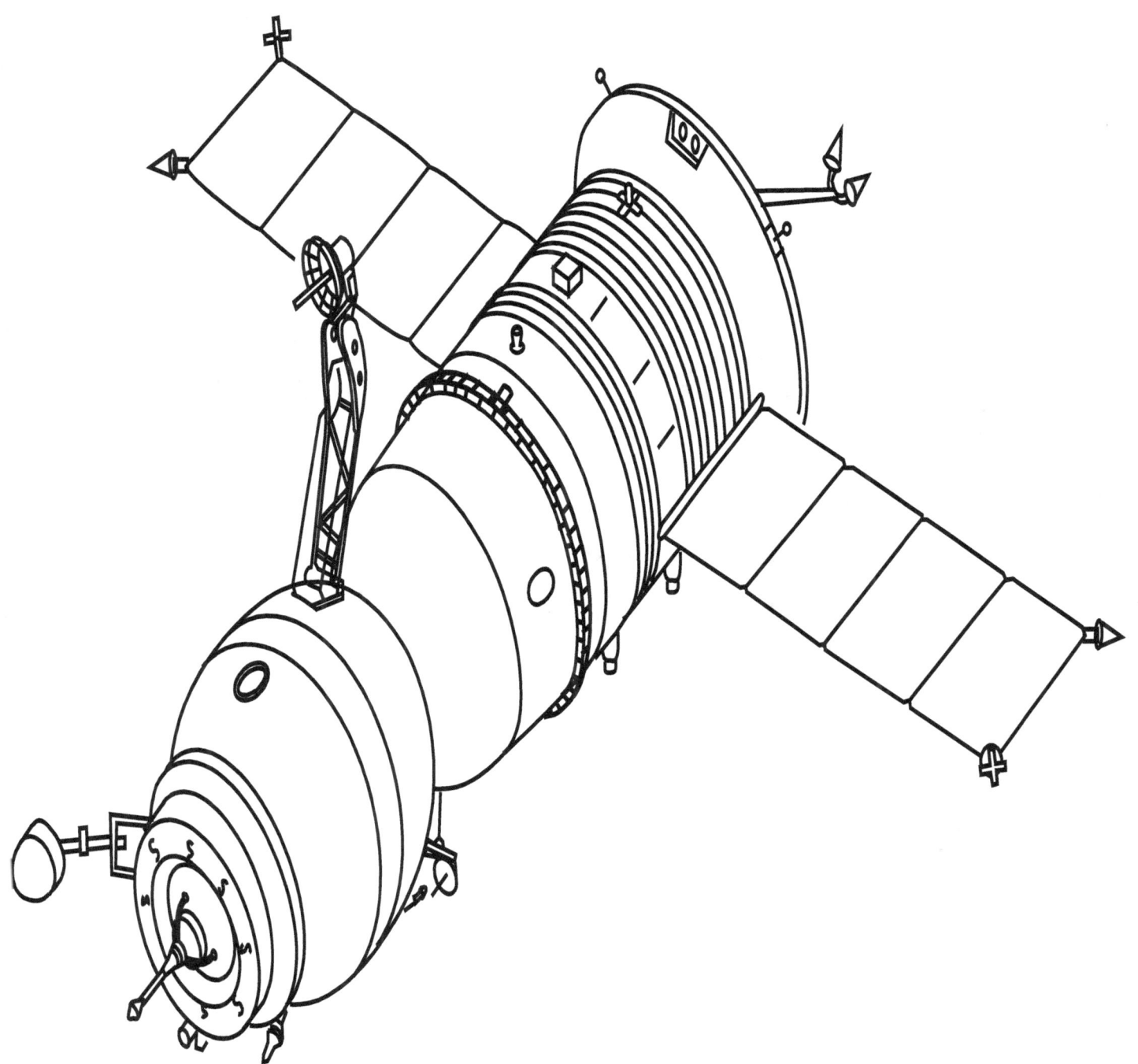

The sunset on Mars appears blue

There are tire tracks and signs of mankind's footsteps on the Moon. The reason for this is because no winds blow on the moon..

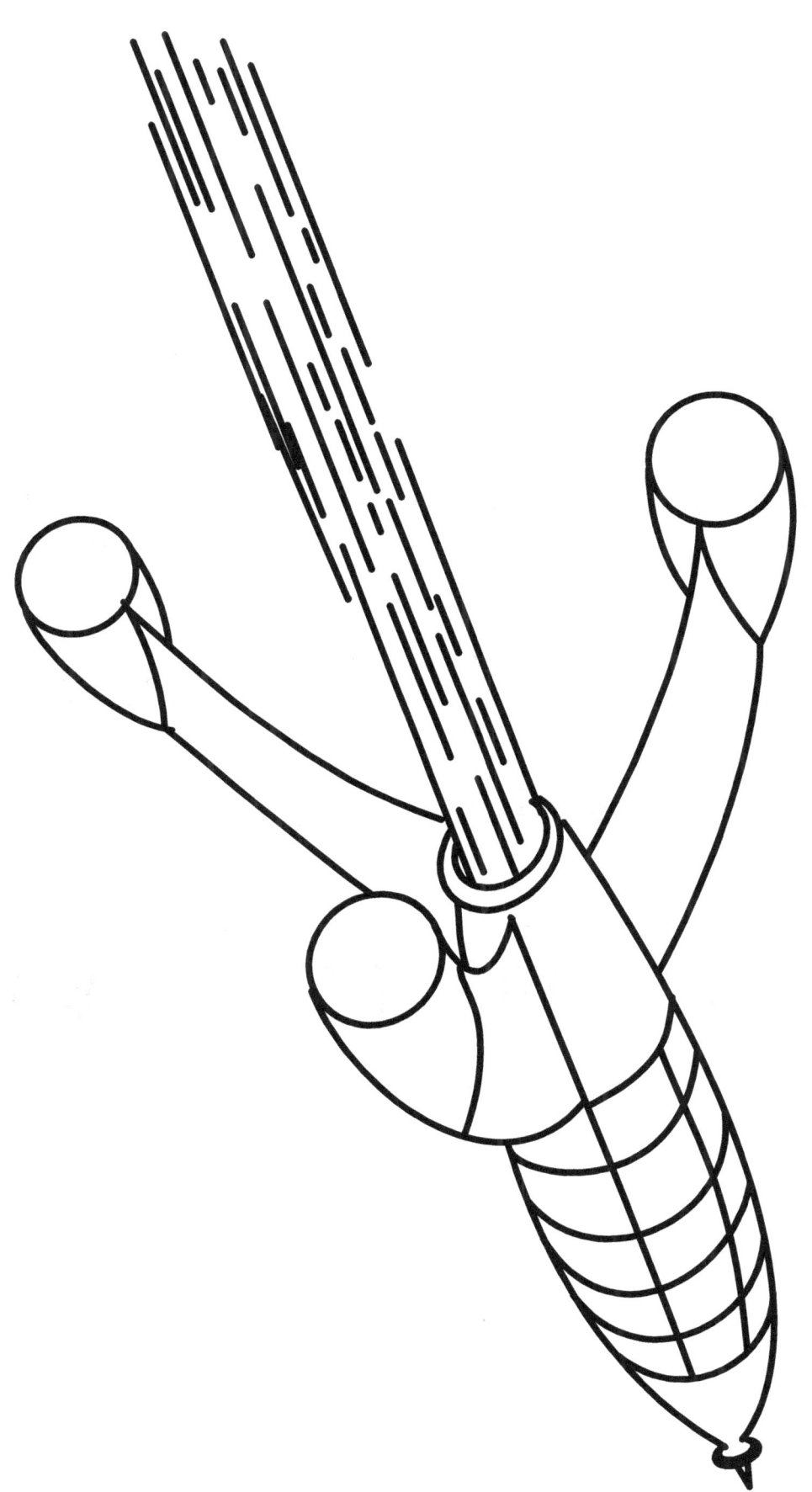

The Earth rotates on an invisible axis and orbits around the sun just like the other seven planets in the Solar System

The high and low tides of oceans are caused due to the gravitational pulls of the moon and the sun.

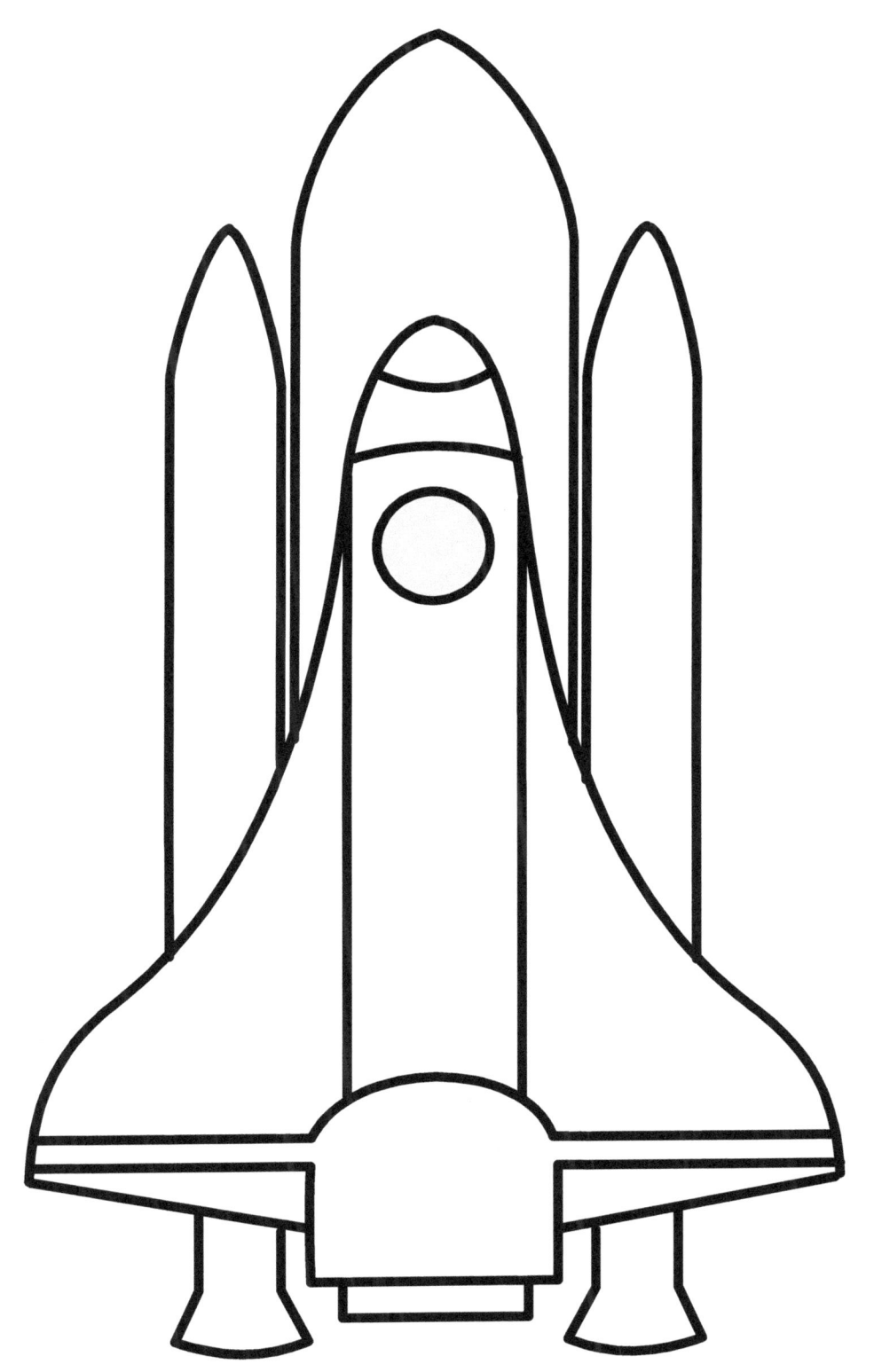

www.ingramcontent.com/pod-product-compliance
Lightning Source LLC
Chambersburg PA
CBHW080618220526
45466CB00010B/3381